Rider-operated lift trucks: Operator training

APPROVED CODE OF
PRACTICE AND GUIDANCE

L117

HSE BOOKS

© Crown copyright 1999
Applications for reproduction should be made in writing to:
Copyright Unit, Her Majesty's Stationery Office,
St Clements House, 2-16 Colegate, Norwich NR3 1BQ

First published 1988, Second edition 1999
Reprinted 2000

ISBN 0 7176 2455 2

All rights reserved. No part of this publication may be reproduced, stored in a retrieval system, or transmitted in any form or by any means (electronic, mechanical, photocopying, recording or otherwise) without the prior written permission of the copyright owner.

This Code has been approved by the Health and Safety Commission, with the consent of the Secretary of State. It gives practical advice on how to comply with the law. If you follow the advice you will be doing enough to comply with the law in respect of those specific matters on which the Code gives advice. You may use alternative methods to those set out in the Code in order to comply with the law.

However, the Code has a special legal status. If you are prosecuted for breach of health and safety law, and it is proved that you did not follow the relevant provisions of the Code, you will need to show that you have complied with the law in some other way or a Court will find you at fault.

This guidance is issued by the Health and Safety Executive (HSE). Following the guidance is not compulsory and you are free to take other action. But if you do follow the guidance you will normally be doing enough to comply with the law. Health and safety inspectors seek to secure compliance with the law and may refer to this guidance as illustrating good practice.

Acknowledgements

HSE gratefully acknowledges the help given by the following organisations.

Barlow Handling
Hereford Fork Trucks (HFT)
JCB Industrial
Lansing Linde

Contents

Notice of Approval *iv*

Preface *v*

PART 1 APPROVED CODE OF PRACTICE — RIDER-OPERATED LIFT TRUCKS: OPERATOR TRAINING

Introduction *1*

What is covered? *1*

Legislation *1*

Obligation to provide basic training *1*

Selection of instructors *2*

Training area and facilities *2*

Training structure and content *2*

Testing *3*

Records *3*

PART 2 GUIDANCE ON GENERAL ASPECTS OF TRAINING

Introduction *3*

Selection of people for training *5*

Training of operators *6*

Basic training *6*

Specific job training *7*

Familiarisation training *8*

Authorisation, records and certificates *8*

Further operator training and monitoring of standards *9*

Instructor selection and training *9*

Recognition of accrediting bodies *10*

Further information *10*

Appendix 1: Recognised accrediting bodies *11*

Appendix 2: Objectives to consider for inclusion in a basic training course *14*

Appendix 3: Basic training: Tests of operator skills *15*

Appendix 4: Example of employer's training record *16*

Appendix 5: Example of an instructional techniques training programme *17*

Further reading *19*

Notice of Approval

By virtue of section 16(4) of the Health and Safety at Work etc Act 1974, and with the consent of the Secretary of State for the Environment, Transport and the Regions, the Health and Safety Commission has on 3 June 1999 approved the revision of the Code of Practice entitled *Rider-operated lift trucks: Operator training*.

The Code of Practice gives practical guidance with respect to the requirements of regulation 9 of the Provision and Use of Work Equipment Regulations 1998 as they relate to the basic training of operators of rider-operated lift trucks.

The Code of Practice comes into effect on 1 October 1999

Signed

ROSEMARY BANNER
Secretary to the Health and Safety Commission

3 June 1999

Preface

Every year there are about 8000 reportable injuries involving lift trucks. These injuries, some fatal, create suffering for those involved and their dependants. They also involve a heavy cost on the employer's business. Even an incident not involving injury may result in costly damage to trucks, buildings, fittings and goods.

Lift-truck accidents are frequently associated with lack of suitable and sufficient operator* training. Such training is an essential first step in reducing damage and injury.

The training of lift-truck operators may be divided into three stages:

Basic training	the basic skills and knowledge required for safe operation;
Specific job training	knowledge of workplace and experience of any special needs and handling attachments;
Familiarisation training	operation on the job under close supervision.

These stages can be taken quite separately, or they may be combined or integrated, particularly where training is carried out on an employer's premises. In either case, it is essential that each stage be covered fully, with due regard to the experience, if any, of the trainees and the type or types of truck they will be expected to operate. The first two stages need to be carried out off the job, ie sheltered from production and other pressures. The third stage needs to be carried out on the job, but under close supervision.

This version of the Approved Code of Practice (ACOP) and guidance is a revision of that first published in 1988. Its coverage has been extended to include a wider range of trucks. The general principles outlined can also be used as a guide in the training of operators of types not covered explicitly.

Part 1, the ACOP, advises on basic training of lift-truck operators. To comply with their duties under the Provision and Use of Work Equipment Regulations 1998 and general duties under the Health and Safety at Work etc Act 1974 employers must ensure that all operators they employ, both new and existing, are adequately trained and, when necessary, provide for their additional or refresher training.

Supplementary advice on basic training is set out in Part 2 (the accompanying guidance). This also gives advice on lift-truck training generally, including specific job and familiarisation training, operator selection and authorisation, and the responsibilities of the self employed and controllers of worksites. The guidance is relevant to all operators. It can serve as a yardstick for the employers of existing lift-truck operators, as well as those newly employed to operate lift trucks.

The ACOP and guidance relate to stacking rider-operated lift trucks, including articulated steering truck types. 'Rider-operated' means any truck capable of carrying an operator and includes trucks controlled from both seated and stand-on positions, which may be fixed or fold-away. Straddle carriers and non lift trucks fitted with removable attachments which modify their function, allowing them to be used temporarily as lift trucks, eg agricultural tractors with fork-lift attachments, are not included.

* For the purpose of this document an operator is anyone who operates a lift truck, even as a secondary or occasional part of their job, and is not limited to people specifically designated as lift-truck operators

Training undertaken in accordance with the ACOP and guidance is not the same as achievement of Scottish/National Vocational Qualifications (S/NVQs). S/NVQs are competence-based qualifications, the award of which denote consistent attainment of defined standards of performance. These qualifications are gained primarily through demonstrated experience in the workplace, and are awarded on the basis of assessments carried out in the workplace. Training may contribute to S/NVQs, but neither is a substitute for the other.

Anyone driving a lift truck on the public highway must comply with the appropriate road traffic legislation.

The ACOP and guidance have been agreed by the Health and Safety Commission following widespread consultation. They give practical advice to help employers meet their legal obligations under the Provision and Use of Work Equipment Regulations 1998 to ensure that all operators receive adequate training for purposes of health and safety.

The ACOP has a special legal status, explained on page (ii). The status of the guidance is also explained on page (ii). Reference in this publication to another document does not imply approval by the Health and Safety Commission of that document except to the extent necessary to give effect to the ACOP and guidance.

For convenience, any text from the Regulations is included in *italic* type.

PART 1 APPROVED CODE OF PRACTICE—RIDER-OPERATED LIFT TRUCKS: OPERATOR TRAINING

Introduction

1 This Approved Code of Practice (ACOP) relates to the provision of basic training for lift-truck operators. It has been produced in consultation with representatives of the Confederation of British Industry, the Trades Union Congress, the Local Authority Associations, the Joint Industry Council (JiC) and others involved with lift trucks.

What is covered?

2 The ACOP covers stacking rider-operated lift trucks, including articulated steering truck types. 'Rider-operated' means any truck capable of carrying an operator and includes trucks controlled from both seated and stand-on positions, which may be fixed or fold-away. The purpose of this coverage is to include all types of lift truck having similar training requirements and to which the advice can reasonably be applied. Straddle carriers and non lift trucks fitted with removable attachments which modify their function, allowing them to be used temporarily as lift trucks, eg agricultural tractors with fork-lift attachments, are not included. Operators of machines adapted for temporary use as lift trucks should be adequately trained to use the attachments they need for the jobs they do.

Legislation

3 Regulation 9 of the Provision and Use of Work Equipment Regulations 1998 (PUWER) requires that:

'(1) Every employer shall ensure that all persons who use work equipment have received adequate training for purposes of health and safety, including training in the methods which may be adopted when using the work equipment, any risks which such use may entail and precautions to be taken.

(2) Every employer shall ensure that any of his employees who supervises or manages the use of work equipment has received adequate training for purposes of health and safety, including training in the methods which may be adopted when using the work equipment, any risks which such use may entail and precautions to be taken.'

4 Regulation 3 of PUWER 1998 extends the requirements of the Regulations to apply to the self employed and those who have control (to the extent that their control allows) of work equipment, people at work who use, supervise or manage the use of work equipment, or the way in which work equipment is used at work.

Obligation to provide basic training

5 Employers should not allow anyone to operate, even on a very occasional basis, lift trucks within the scope of this ACOP who have not satisfactorily completed basic training and testing as described in this ACOP, except for those undergoing such training under adequate supervision.

ACOP

Selection of instructors

6   When arranging for training, employers should satisfy themselves that it is in accordance with this ACOP. Operator training should only be carried out by instructors who have themselves undergone appropriate training in instructional techniques and skills' assessment. They should give instruction only on the types of lift truck and attachments for which they have been trained and successfully tested as operators. Instructors also need sufficient industrial experience to enable them to put their instruction in context and an adequate knowledge of the working environment in which the trainee will be expected to operate.

Training area and facilities

7   Basic training may be given at a suitable training centre or venue, or on an employer's premises. Where practicable, training areas should be sheltered from adverse weather conditions.

8   Basic training needs to be carried out off the job. Even when conducted on an employer's premises this means that the instructor and trainees, together with the lift truck and loads, should be wholly concerned with training, kept away from normal commercial operations, and not be diverted to other activities while training is in progress. Lift trucks used for training must be in good mechanical condition, properly maintained (taking into account manufacturers' recommendations), conform to all legal requirements and be suitable for the particular uses to which they will be put.

9   A suitable manoeuvring area should be provided and appropriately marked. While training is in progress access to this area should be restricted to the instructor and trainees. The area will need to include facilities for simulating the manoeuvring space likely to be encountered in the workplace, including slopes. For rough terrain trucks an appropriate surface and obstacles representative of the conditions for which training is being provided is necessary.

10   A supply of realistic loads appropriate to the training being given, such as loaded and unloaded pallets, bags, sacks, bales, drums, bulk materials and freight containers is necessary to make training realistic. Similarly, there should be appropriate facilities for simulating loading and unloading from racking at various heights as well as road vehicles.

11   A training room or other suitable accommodation, together with appropriate training aids (eg projectors, models) should be made available to enable the instructor to cover, under reasonable conditions, the principles of lift-truck operation.

Training structure and content

12   Training should be largely practical in nature and of sufficient length to enable trainees to acquire the basic skills and knowledge required for safe operation, including knowledge of the risks arising from lift-truck operations. It should not be altered to suit immediate operational or production needs.

13   The ratio of trainees to instructors needs to allow each trainee adequate time to practise operating the truck under close supervision and to prepare for the practical tests.

ACOP

14 Training should follow a carefully devised programme which ensures that each stage is introduced in an appropriate sequence, building on what has gone before, and allowing adequate time for learning and practice before the next stage is tackled. The easier driving skills should be dealt with before progressing to more difficult operations such as pallet or other load handling. At each stage the instructor will need to explain and demonstrate safe operation, which should then be practised by the trainees under direct supervision.

15 Basic training should be given on all the types of lift truck and attachments that operators will or could be required to use in their work. If the operator is subsequently required to operate another type of lift truck, or there is a change of handling attachment, additional, practical conversion training will be required. Employers should also consider the need for conversion training where the truck type does not change, but the size and weight alters significantly.

16 The course content will depend upon the lift-truck operations the trainee will be expected to carry out. The objectives of a basic training course, some of which are listed in Appendix 2, need to be tailored to fit all the lift-truck operations to be undertaken by the operator.

Testing

17 The instructor should assess a trainee's progress continuously to ensure that the required standards are achieved at each stage of basic training. Additionally, trainees are required to pass a test or tests, practical and theoretical, of the skills and knowledge needed for safe operation.

Records

18 Employers need to keep a record for each employee who has satisfactorily completed basic training and testing in accordance with this ACOP. The record should include sufficient information to identify the employee and the nature and content of the training and testing completed. Either a copy of any certificate of basic training issued, or the relevant details, should be included in employers' records.

PART 2 GUIDANCE ON GENERAL ASPECTS OF TRAINING

Introduction

Guidance

19 This guidance supplements the Health and Safety Commission's Approved Code of Practice (ACOP) on the basic training of lift-truck operators. Like the ACOP, it relates to stacking rider-operated lift trucks (which includes articulated steering truck types), and excludes straddle carriers and non lift trucks fitted with removable attachments which modify their function, allowing them to be used temporarily as lift trucks, such as agricultural tractors with fork-lift attachments. 'Rider-operated' means any truck capable of carrying an operator and includes trucks controlled from both seated and stand-on positions, which may be fixed or fold-away. This does not mean that training only needs to be given for these lift trucks. The employer's duty under the Provision and Use of Work Equipment Regulations 1998 to provide training extends to operators of all other types of truck. The advice given in the ACOP and guidance can be used as an indication of the standard of training to provide for all types of lift truck. The guidance can be of help not only to employers, but also to organisations offering training for operators and instructors, and to lift-truck suppliers.

Guidance

20 Safe operation of any plant or machinery requires proper training. It is quite wrong to assume that because employees hold a licence to drive, say, a motor vehicle on the public roads, they also have the skills necessary to operate a lift truck.

21 Employers are responsible for ensuring that adequate training is provided for their employees. Employers should satisfy themselves that any training given covers all aspects of the work to be undertaken and takes account of this guidance, and in the case of basic training, is at least to the standard of the ACOP. Self-employed lift-truck operators also have responsibilities under the Provision and Use of Work Equipment Regulations 1998 to ensure they undergo the same type of training, achieving the same standard, as employers are required to provide to their employees.

22 **Operators of types of lift truck not covered by this guidance will also need training.** In some cases, for instance pedestrian/rider-operated pallet trucks, this may follow a similar approach, but in others, such as straddle carriers, a very different training programme will be needed. Training organisations involved in lift-truck training should be able to advise on suitable training, but employers will need to take account of the advice on instructor selection contained in this guidance when choosing a training provider, to ensure that the provider has the relevant expertise and experience. The bodies listed in Appendix 1 are able to advise about training providers.

23 Employers also have a continuing responsibility to provide adequate supervision and it is therefore essential that supervisors themselves have sufficient training and knowledge to recognise safe and unsafe practices. This does not mean that supervisors need full operator training, but they do need to understand the risks involved, and the means of avoiding or counteracting them. Training in health and safety management, risk assessment and safe systems of work should be considered. Advice on risk assessment can be found in the following publications by HSE: *Management for health and safety at work* L21 and *5 steps to risk assessment* INDG163 (see Further reading for details). Supervisors also need sufficient training to enable them to evaluate the advice of fully trained and experienced operators to ensure they do not over-ride the operator's advice and reduce safety.

24 Safety representatives appointed in accordance with the Safety Representatives and Safety Committees Regulations 1977 should be consulted about the training arrangements for lift-truck operators. If there are no appointed safety representatives, employers will need to consult with their employees either through elected safety representatives or directly in accordance with the Health and Safety (Consultation with Employees) Regulations 1996. Safety representatives and employees can play an important role in encouraging the safe operation of lift trucks.

25 Employers should ensure that employees (eg lorry drivers, maintenance or inspection personnel) who use lift trucks on other people's premises are fully trained to do so, and that information to this effect is made available to controllers of those premises. Information provided could be documentation on an individual basis or written assurance that all their employees who will visit the site and be expected to operate lift trucks are trained and competent to do so. Site controllers should use this information to satisfy themselves, before allowing use of trucks, that visitors have been adequately trained to safely operate the lift truck(s) to be used. Site controllers also need to provide visitors with site specific information, which could be provided by clear signs or, where risk assessment indicates the need, site vehicle and third party rules, to enable them to work safely. A useful precaution might be to limit clearly

Guidance

areas where people who are not familiar with the premises are allowed to operate. It is highly unlikely that visiting lorry drivers will have undergone training which would enable them to safely use lift trucks provided by occupiers of worksites. Drivers with their own lift trucks or regular contractors' drivers who frequently visit the same sites may be satisfactorily trained and have sufficient site knowledge to operate safely.

26   Employers who do not control worksites where their employees may operate lift trucks and those who do control such sites need to co-ordinate their efforts and co-operate to ensure that only people trained as described in this ACOP and guidance are allowed to operate lift trucks. Such co-ordination and co-operation is equally important on multi-occupied sites, such as business parks and markets where trucks may be shared. The responsibility of those who control worksites to ensure that the workplace is safe in no way detracts from the employer's duty to ensure that their own employees are adequately trained.

27   Employees also have responsibilities. Section 7 of the Health and Safety at Work etc Act 1974 requires them to take reasonable care for their own health and safety and that of other people. They must also co-operate with their employers to assist them in complying with their statutory duties. Section 8 of the Health and Safety at Work etc Act 1974 requires that employees should not interfere with or misuse anything provided in the interests of health, safety or welfare under health and safety legislation.

Selection of people for training

28   Employers should select potential lift-truck operators carefully. Those selected for training need to have the ability to do the job in a responsible manner and the potential to become competent operators. Operators of lift trucks on docks premises must be aged at least 18 years. For advice on the Docks Regulations 1988 see *Safety in docks* (details in Further reading section). Young persons (under 18 years of age) are often exposed to risks to their health and safety when using work equipment as a consequence of their immaturity, lack of experience or absence of awareness of existing or potential risks. Therefore, such young people should not be allowed to operate lift trucks without adequate supervision unless they have the necessary competence and maturity, as well as having successfully completed appropriate training. For more information see The Health and Safety (Young Persons) Regulations 1997 (SI 1997 No 135) which applies to young workers aged under 18 years and *Young people at work - A guide for employers* which gives guidance on the Regulations. Children under minimum school leaving age should never operate lift trucks.

29   Those selected should have the necessary level of physical and mental fitness and learning ability for the task. People with disabilities may well be able to work safely with lift trucks. In cases where a disability is potentially relevant to the safe operation of lift trucks, employers should seek medical advice on a case by case basis. The Disability Discrimination Act 1995 may apply. For an explanation of employers' duties under this Act see the Department for Education and Employment's Code of Practice (details in Further reading section). Further information on medical considerations is given in HSE's booklet HSG6 *Safety in working with lift trucks* (see Further reading section).

30   Where employees claim to be trained and experienced, employers should insist upon evidence. Employers need to satisfy themselves that the training, experience and ability is in fact sufficient and relevant to the lift trucks and handling attachments to be used. Where evidence, such as a training

Guidance

certificate, is not available, employers will need to arrange assessment of the person's competence and provide any training which the assessment indicates is necessary before allowing the employee to operate a lift truck.

31   It may be useful to apply a selection test to avoid wasteful attempts to instruct unsuitable trainees. Advice on trainability assessment can be obtained from the bodies listed in Appendix 1.

Training of operators

32   The training of operators should always include the three stages of training:

Basic training	the basic skills and knowledge required for safe operation;
Specific job training	knowledge of workplace and experience of any special needs and specific handling attachments;
Familiarisation training	operation on the job under close supervision.

The first two stages are sometimes combined or integrated but should always be off the job. The ACOP covers basic training but further guidance on this is included below.

Basic training

33   Basic training needs to cover fully the skills and knowledge required for the safe operation of the type of lift truck and handling attachments (if any) which the trainee will be required to operate, including the risks arising from lift-truck operation. Such risks would include not only those directly related to the operation of trucks, but associated tasks, such as the fire hazard created by possible production of hydrogen when recharging batteries. Appendix 2 lists objectives which may need to be included.

34   Length of training may vary depending on the objectives to be covered, the trainee/instructor ratio and the ability and previous experience of the trainees. For instance, an agricultural tractor driver may need less training on a rough terrain truck than a complete novice. As a rough guide, the normal length of a course for novice operators would be five days. In all cases, the time devoted to training needs to be sufficient to ensure that the basic training objectives can be achieved.

35   Operators with some experience of lift trucks or relevant experience of similar vehicles may need less extensive training than those with no experience. **However, the value of such experience should not be overestimated**. The ability to drive private cars or other conventional road vehicles, for example, does not remove the need for proper training on lift trucks, which have very different stability and handling characteristics as well as different controls. An operator with basic training on one type of lift truck or handling attachment cannot safely operate others, for which they have not been trained, without additional, conversion training.

36   Training providers can arrange short assessment courses to judge the ability and training needs of experienced operators who have had limited formal training.

Guidance

37   Given the wide range of lift trucks, operator experience and company requirements, some training organisations will arrange for a basic course to be tailored to meet a client's requirements. The basic training described in the ACOP can be adapted for this purpose, provided always that the appropriate basic training objectives are achieved.

38   It is beyond the scope of this guidance to give detailed examples of suitable courses. Appendix 1 lists five bodies which operate accreditation schemes for the training, testing and certification of operators; a brief description of their schemes is included. Further information can be obtained from the bodies themselves.

39   The ratio of trainee : instructor : truck should enable the instructor to demonstrate each part of the practical training and the trainee to obtain adequate hands-on experience. There should be adequate time for each trainee to have sufficient practical experience to become a safe operator and to do so under close supervision. A trainee : instructor : truck ratio of 2:1:1 is probably ideal, but in any case the ratio should not exceed 3:1:1 except for lecture sessions. The opportunity to learn from the performance of other trainees can be valuable.

40   Trainees need to be continually assessed. Appendix 3 lists tests of operator skills which may be appropriate. Advice on operator testing may be obtained from the bodies listed in Appendix 1.

41   It is essential that newly trained operators be given specific job and familiarisation training as described below. Once fully trained, operators should also be given the opportunity to put the skills and knowledge acquired during training into practice at their workplace to reinforce that training. Newly acquired skills can swiftly be lost if not used.

Specific job training

42   Specific job training is a further essential element of training. It will normally follow the completion of basic training but may be combined or integrated with it. The trainee : instuctor : truck ratio for basic training also applies to specific job training.

43   Specific job training will be tailored to the employer's special needs and include, where appropriate:

(a)   knowledge of the operating principles and controls of the lift truck to be used, especially where these relate to handling attachments specific to the job, or where the controls differ from those on which the operator has been trained. Routine inspection and servicing of that truck in accordance with the operator's handbook or instructions issued by the manufacturer need to be covered, in so far as they may reasonably be carried out by the operator. This should be repeated whenever the design of truck is changed;

(b)   use of the truck in conditions that the operator will meet at work, eg gangways; loading bays; racking; lifts; automatic doors; confined areas; cold stores; slopes; rough terrain; loading platforms; and bad weather conditions;

(c)   instruction on site rules, eg site layout; one-way systems; speed limits; general emergency procedures; use of protective clothing and devices

Guidance

including operator restraints and eye and hearing protection; work near excavations, overhead lines and other hazards;

(d) training in the work to be undertaken, eg loading particular kinds of vehicle; handling loads and materials of the kind normally found at that workplace, including assessment of weight; use of the fork truck to support working platforms where appropriate; for advice on the safe use of lift trucks for this purpose, see HSE's guidance note PM28 *Working platforms on fork-lift trucks* (see Further reading section); and

(e) safe systems of work, which should include custody arrangements to ensure that keys are never left in unattended trucks, or in a place where they are freely available, to prevent the use of trucks by unauthorised operators.

Familiarisation training

44 This is the third element of training. It needs to be carried out on the job and under close supervision, by someone with appropriate knowledge, possibly the trainee's usual supervisor. It should cover the application, under normal working conditions, of the skills already learned and include familiarisation with site layout, local emergency procedures and any other particular feature of the work which it is not practicable to teach off the job. In **very** exceptional circumstances, such as use of lift trucks by the emergency services at the scene of an accident or fire, where it is clearly not feasible to train on-site, realistic simulated training may be provided.

Authorisation, records and certificates

45 Following satisfactory completion of training, the employee should be given written authorisation to operate the type or types of truck for which all three elements of training have been successfully completed. Authorisations may be issued on an individual basis and/or recorded centrally by the employer. Authorisations should state the operator's name, the date of authorisation, the truck or trucks to which they relate and any special conditions, such as area limitations. Employers should not allow personnel to operate lift trucks on any premises without authorisation (except in the case of a trainee under close supervision). Employers will also need to ensure that they are satisfied with the continuing competence of authorised operators.

46 Employers should keep adequate records for each employee who has satisfactorily completed any stage of lift-truck training. The record will need to include sufficient information to identify the employee and the nature of training completed. An example of an employer's training record is given in Appendix 4. The record could include a copy, or details, of any certificate of training which is issued. While there is no legal requirement for certificates of basic training to be issued, they are strongly recommended as a useful, practical means of providing documentary evidence of relevant training having taken place and an appropriate level of operating ability having been attained. The employee will need a certificate as evidence of training on any change of employment. It is in the interests of both employers and employees for employees to have the original certificate to limit the opportunities for forgery which photocopies present. If only a copy is provided to the employee it will need to be annotated in some way to establish its validity for the purpose of recognition by other employers.

> **Guidance**

Further operator training and monitoring of standards

47 There is no specific requirement to provide refresher training after set intervals, but even trained and experienced lift-truck operators need to be re-assessed from time to time to ensure that they continue to operate lift trucks safely. This assessment, which should form part of a firm's normal monitoring procedure and be formally time-tabled to ensure that it is done at reasonable intervals, will indicate whether any further training is needed. In addition to routine safety monitoring, re-assessment might be appropriate where operators have not used trucks for some time, are occasional users, appear to have developed unsafe working practices, have had an accident or near miss, or there is a change in their working practices or environment. Employers may find it useful to record re-assessment in their safety monitoring records. Employers can, of course, decide that automatic retraining after a set period of time is the best way of ensuring that employees are adequately trained but, where this approach is adopted, it will still be necessary to monitor performance in case retraining is required before the set period ends. The guiding principle is that employers need to maintain the competence of operators to use lift trucks safely through a laid down, formal process of monitoring and assessment.

48 Conversion training, to enable operators to extend the range of trucks they are qualified to drive, may also be appropriate and is widely available. Refresher and conversion training should be approached with the same attention to detail as basic training to ensure that all gaps in and variants on existing skills and knowledge are identified and covered during training. For instance, there may be significant variations in the arrangement or application of controls, even in the same truck types.

49 Training will not in itself ensure the competence of individuals: this will develop with experience and should be monitored. Continued supervision will be necessary to ensure that good standards of operation are maintained.

Instructor selection and training

50 Successful training depends on the competence of instructors. They should be asked to supply evidence of their training and post-training experience on the type of truck to be used, both as instructor and operator, and their knowledge of and familiarity with conditions in the industry where trainees will work. This will include expertise in any requirements peculiar to the operation of the truck(s) and in the work trainees will be expected to undertake. Since training is largely accomplished through demonstration followed by supervised practice, it is essential that each demonstration by the instructor is a model, free from technical errors and misjudgements. Instructors must also be able to make effective use of instructional techniques in both the working and classroom environment.

51 Good instructors should:

(a) have the ability to adapt their approach to suit the needs of different trainees;

(b) be able to communicate effectively;

> **Guidance**

(c) be able to lead and control; and

(d) keep their own training and experience as instructors up to date, especially if not training regularly.

52 An example of an instructional techniques training programme is in Appendix 5. Instructors need to be re-assessed periodically as appropriate. Such re-assessment is particularly important if instructors have not done any training for some time. Advice on instructor training can be obtained from the bodies listed in Appendix 1.

Recognition of accrediting bodies

53 The Health and Safety Commission has recognised the five bodies listed in Appendix 1 as competent to operate voluntary accreditation schemes. Such schemes are not mandatory but recognition by the Commission is intended to help set and maintain professional training standards. This should help employers to select training organisations or lift-truck suppliers who offer a good standard of training. Although employers operating their own in-house training schemes may also find it useful to have them accredited by one of the above bodies, this may not be necessary for successful schemes which operate in some companies.

Further information

36 The Further reading section lists additional sources of information on lift-truck training and related subjects.

Appendix 1 Recognised accrediting bodies

1 The bodies listed below have been recognised by the Health and Safety Commission as competent to accredit and monitor organisations to train instructors and/or to train, test and certificate operators. The nature and scope of their accreditation schemes are briefly described.

Association of Industrial Truck Trainers (AITT), Huntingdon House, 87 Market Street, Ashby de la Zouch, Leicestershire, LE65 1AH (Tel: 01530 417234).

2 The association has established the Independent Training Standards Scheme and Register (ITSSAR) as a system of examination and registration of lift-truck instructors, examiners and tutors (trainers of instructors), and monitoring of operator training for all truck types, across all industry sectors, which is available to both association members and non-members alike. The association offers accreditation to:

(a) instructors;

(b) examiners;

(c) tutors;

(d) organisations conducting lift-truck operator courses, including employers' in-house schemes;

(e) organisations conducting both lift-truck instructor and operator training courses.

3 ITSSAR monitors the examination of lift-truck instructors, issues personal identity cards and certificates and keeps a register for each of the above categories. Registration under categories (a), (b) and (c) is for five years and instructors and examiners may be re-registered after retaking and passing the examination. Tutors are progressively examined. Accreditation of organisations is valid for one year and is renewable subject to their meeting the requirements of the association's published conditions.

4 It also offers the facility for ITSSAR instructors and organisations to register operators on a national registration scheme.

5 Further details can be obtained from the Independent Training Standards Scheme and Register, Scammell House, High Street, Ascot, Berkshire SL5 7JF (Tel: 01344 874454).

Construction Industry Training Board, Bircham Newton, King's Lynn, Norfolk PE31 6RH (CTA national helpline Tel: 01485 577838).

6 The board administers a Certificate of Training Achievement (CTA) scheme for construction plant operators on behalf of the Construction Plant-hire Association and the Construction Confederation. The scheme was introduced in July 1986 and includes separate categories to cover 43 different items of construction plant. These include rough terrain, industrial counterbalanced, reach and side loader, and aligns with the Intermediate Construction Certificate (ICC) which is part of the NVQ framework.

7   For the purpose of the scheme, the board:

(a) accredits training organisations and instructors;

(b) approves operator training, which may take place in-company or at a training centre, through accredited organisations;

(c) issues operator/instructor certificates of training achievement and construction certificates following assessment by CTA accredited instructors;

(d) maintains a register of certificated plant operators, accredited instructors and accredited training organisations.

Further details can be obtained from the CTA unit, CITB, Bircham Newton.

Lantra National Training Organisation Ltd, NAC, Kenilworth, Warwickshire CV8 2UG (information services helpline, Tel: 0345 078007).

8   Lantra National Training Organisation Ltd provides access to training for those working in land-based and related industries. This includes a national instructor registration scheme, a national training provider scheme and access to operator training. Training covers lift trucks of the types used in those industries including industrial and rough terrain counterbalanced lift trucks and telescopic materials handlers, and the range of commonly used attachments.

9   Lantra's scheme:

(a) provides access to operator training courses, usually run on employers' premises;

(b) approves operator training providers and centres;

(c) provides certificates of training achievement when operators have met training objectives;

(d) trains, assesses, certificates and registers instructors.

Further details can be obtained from Lantra National Training Organisation Ltd on its information services helpline, Tel: 0345 078007.

National Plant Operators Registration Scheme Ltd, Highfield Farm, Lostock Gralam, Cheshire CW9 7PL (Tel: 01606 49909).

10 The scheme provides:

(a) a national register of qualified instructors who have successfully completed a methods of instruction course and have proven ability to both operate and train on the type(s) of lift truck on their remit;

(b) a national register of qualified operators of industrial and rough terrain lift trucks, and telescopic materials handlers;

(c) a system for approving operator training courses conducted at approved training centres or in-company;

(d) a system for approving training centres for both operator and instructor training courses; and

(e)   training courses for lift-truck instructors.

RTITB Ltd, Training, accreditation and examiner services,
Ercall House, 8 Pearson Road, Central Park, Telford TF2 9TX
(Tel: 01952 520200; Fax: 01952 520201).

11   RTITB Ltd (formerly the Road Transport Industry Training Board) offers accreditation of lift-truck operator and instructor training to:

(a)   commercial training organisations providing operator training at permanent centres or on customers' premises with suitable facilities;

(b)   employers' own in-house lift-truck training schemes;

(c)   commercial training organisations providing lift-truck instructor courses.

12   Accreditation is available to any individual or organisation able to meet and continue to comply with RTITB published standards. It is valid for a 12 month period, subject to monitoring visits by RTITB assessors, and is renewable provided standards are maintained. The RTITB accreditation scheme can be applied to training carried out on any truck type and in any commercial or industrial environment.

13   After successful completion of training and assessment by their trainer, instructors are independently examined by an RTITB examiner. After passing the examination, instructors become eligible for inclusion on the RTITB National Register of Qualified Instructors and are issued with a personal identification badge and certificate. Qualified and registered instructors may then apply to become RTITB accredited training providers in their own right.

14   Further information and advice on any of its schemes and services may be obtained from RTITB Ltd at the address, telephone and fax numbers given.

Appendix 2

Objectives to consider for inclusion in a basic training course

On completion of training, the trainee should be able to do the following.

1   State the reasons for operator training, the risks associated with lift-truck operations and the causes of lift-truck accidents.

2   State the responsibilities of operators to themselves and others, including their duties under sections 7 and 8 of the Health and Safety at Work etc Act 1974.

3   Identify the basic construction and main components of the lift truck, stating its principles of operation and load handling capabilities and capacities.

4   Identify, as appropriate, handling attachments which may be used with the lift truck.

5   Locate and state the purpose and method of use of all controls and instruments.

6   Place the forks or other handling attachment in pre-determined positions employing the appropriate controls.

7   Identify various forms of load, and state the procedures for their stacking, destacking and separation; assess the weight, and, where relevant, the load centre of a load; and decide if the load with its known weight and load centre is within the truck's rated/derated capacity.

8   State the factors which affect machine stability, including: turning, especially related to speed and sharpness of turn; load security and integrity; rated capacity and rated load centres; centres of gravity; and speed and smoothness of operation.

9   Follow correct procedures when loading and unloading vehicles.

10   Make visual checks to ascertain the safety, soundness and rating of structures designed to receive loads, and place and remove loads on and from those structures at various heights.

11   Pick up and place loads, and drive and manoeuvre the machine in forward and reverse motions laden and unladen on inclines, in restricted spaces and on level ground (including rough terrain as applicable) following correct procedures and precautions.

12   Park the machine, following correct procedures and precautions;

13   where applicable state the purpose, and demonstrate the procedures for the use, of safety devices including stabilisers, level indicators, and load indicators, if fitted.

14   Carry out inspection and maintenance tasks appropriate to operators as required by the machine manufacturers and any relevant legislation.

15   State the actions to be taken in an emergency while in control of a lift truck, for example, action to be taken in the event of tipover.

16   State why it is essential to have vehicle key custody arrangements.

Appendix 3   Basic training: Tests of operator skills

Testing could include the following as appropriate.

1   Operation of the truck within the safety limits defined by the manufacturer.

2   Carrying out a pre-use check when the truck is to be used.

3   Correct mounting and dismounting procedure and correct driving position.

4   Competent use of controls.

5   Movement of the truck with forks or attachments in the correct travel position, laden and unladen.

6   Correct insertion and withdrawal of forks or other handling attachments without damage to pallet or load.

7   Manoeuvring a loaded truck forward and in reverse in a narrowly confined area.

8   Performing both a left and a right 90° turn with a loaded truck in a narrowly confined area without touching the sides of the area.

9   Stacking and destacking loads:

(a)   at different levels;

(b)   in front of a fixed vertical face;

(c)   on the floor alongside similar loads.

10   Loading/unloading a vehicle (a suitable simulation may be used where a vehicle is not available).

11   Correct parking of the truck.

Appendix 4

Example of employer's training record

Company name:
Company address:

Employee's full name: Department:
Clock number: NI number:

Basic training

Lift truck type(s) used for training:
Model/capacity:
Attachments:
Organisation carrying out training:
Course description, location and reference number:

Duration and dates of course: days from   to
Name of instructor:
Reference number:
Date of test:
Name of person conducting test:
Reference number:

Specific job training

Lift truck(s) used for training:
Model:
Number:
Instructed by:
Duration of training:
Date of training:

Familiarisation training

Lift truck(s) used for training
Model:
Number:
Instructed by:
Duration of training:
Date of training:

Appendix 5 — Example of an instructional techniques training programme

The following outline programme for a lift-truck instructor course is given as an example. Full details of instructor training courses can be obtained from the bodies listed in Appendix 1.

Course aims

To provide effective instruction on the following:

(a) relevant instructional skills, including classroom technique;

(b) techniques for structuring training material into a logical sequence;

(c) an objective and critical approach towards the effectiveness of the instruction presented;

(d) appropriate methods of assessment on the progress of trainees and the testing of basic skills and knowledge they acquire.

Course objectives

On completion of training, course members should be able to demonstrate their ability to plan, prepare and present practical and theoretical instruction to an adequate standard. In addition, they should be able to construct, conduct and mark objectively a practical test or tests of trainees' operating ability and issue appropriate documentation.

Course content

(a) Principles of instruction;

(b) Simple task analysis;

(c) Preparing a job break-down sheet;

(d) Planning, preparing and presenting a practical demonstration;

(e) Planning, preparing and presenting a practical lesson;

(f) Planning, preparing and presenting a classroom lesson;

(g) Use of question and answer techniques;

(h) Developing training courses suitable for new trainees or experienced operators;

(i) Guidance in specific job and familiarisation training;

(j) Constructing, conducting and marking practical and theoretical tests of trainees' operating ability;

(k) Certification of operators after basic training;

(l) Employers' authorisations to operate.

Final instructor assessment

Course members should be assessed on:

(a) practical operating and instructing ability;

(b) theoretical instructing ability;

(c) ability to conduct and mark tests of trainees' operating ability and knowledge to an appropriate and consistent standard.

Further reading

Safe use of work equipment. Provision and Use of Work Equipment Regulations 1998 ACOP and guidance L22 HSE Books 1998 ISBN 0 7176 1626 6

Safe use of lifting equipment. Lifting Operations and Lifting Equipment Regulations 1998 ACOP and guidance L113 HSE Books 1998 ISBN 0 7176 1628 2

Management of health and safety at work. Management of Health and Safety at Work Regulations 1992 ACOP L21 HSE Books 2000 ISBN 0 7176 2488 9

5 steps to risk assessment INDG163 (Rev1) HSE Books 1998

Safety in working with lift trucks HSG6 HSE Books 1993 ISBN 0 1716 1440 9 Currently under revision

Safety in docks. Docks Regulations 1988 ACOP with regulations and guidance COP25 HSE Books 1988 ISBN 0 7176 1408 5

Workplace transport safety: Guidance for employers HSG136 HSE Books 1995 ISBN 0 7176 0935 9

Young people at work: A guide for employers HSEG165 HSE Books 1997 ISBN 0 7176 1285 6

Preventing accidents to children in agriculture ACOP L116 HSE Books 1999 ISBN 0 7176 1690 8

Working platforms on fork-lift trucks PM28 HSE Books 2000 ISBN 0 7176 1233 3

A guide to information, instruction and training: Common provisions in health and safety law INDG235 HSE Books 1996

Safe use of vehicles on construction sites HSG144 HSE Books 1998 ISBN 0 7176 1610 X

Other publications

Code of Practice for the elimination of discrimination in the field of employment against disabled persons or persons who have had a disability Department for Education and Employment, HMSO 1996 ISBN 0 11 270954 0. Available from The Stationery Office, Publications Centre, PO Box 276, London SW8 5DT (Tel: 0870 600 5522) and from The Stationery Office Bookshops

Scheme for the certification of training achievement of construction plant operators Construction Industry Training Board 1995 (address in Appendix 1)

Instructional techniques and methods - Instructor guides and *Training and testing standards for the whole range of lift trucks* Independent Training Standards Scheme and Register (address in Appendix 1)

RTITB recommendations for the training of lift truck operators and instructors; RTITB accreditation criteria for lift truck training RTITB Ltd (address in Appendix 1)

Recommendations for the safe operation of large lift trucks and ro-ro terminals 1983 Available from PSO (Technical Services) Ltd, 220 Africa House, 64-78 Kingsway, London WC2B 6AH (Tel: 0171 242 3538)

Industrial trucks on public roads GN57 1998 Available from the British Industrial Truck Association, Scammell House, 9 High Street, Ascot, Berkshire SL5 7JF (Tel: 01344 623800)

While every effort has been made to ensure the accuracy of the references listed in this publication, their future availability cannot be guaranteed.